Pattern Designs by Sheeraz

50 Pattern's Sketches and 200 Colored Patterns

Pattern Design by Sheeraz

First Edition

For:

Architecture Massing | Architecture Surface Design

Decoration | Textile Design | Art & Graphics |

Branding | Logo Design | Packaging Design |

Children Stuff because patterns attract children

Published by
Shamsi Publisher
COES&RJ LLC.,
Centre of Excellence for Scientific & Research Journalism
10685-B Hazelhurst Dr. # 16258,
Houston, TX- 77043, USA.
Phone: +1-281-407-7509
Fax   : +1-281-754-4941
Email: info@centreofexcellence.net
Website: www.centreofexcellence.net
Houston – London – Karachi – Singapore
Title Design by: Muhammad Sheeraz

Copyright © Muhammad Sheeraz, 2021
All rights reserved
Page 146 constitutes an extension of this copyright page.

ISBN: 978-1-257-76636-9

Without limiting the rights under copyright
reserved above, no part of this publication
may be reproduced, stored in or introduced into
a retrieval system, or transmitted, in any form
or by any means (electronic, mechanical, photocopying,
recording or otherwise), without the prior written
permission of the creator of Pattern Designs of this book.

To:

Mr. Adnan Asdar Ali and Mr. Ariff Ul Islam

www.aaaprojects.com

and

Everyone who likes pattern designs

The pattern designs in this book are, as continuation of
"Attainable Housing Standards by Sheeraz"
Available in the market and online bookstores

Standard No. 8 (Architectural Design with Parking)
for residence needs without compromising the regulations

and

Standard No. 9 (Design & Color Customization)
for architectural design & amendment by customization

To,

Provide the surface design options to make the house feel like a home.

DISCLAIMER

All the patterns introduced herein, specifically designed for this book. Not a single pattern copied from anywhere. Some pattern designs may look like similar to the already available pattern designs in the field, but the patterns in line works and then creating by color shades, are distinguishing them with the others, which are already available.

Some mistakes or errors in terms of design and presentation may be found here. All pattern's sketches and colored patterns herein not presented with an intent to harm to any person and organization or substitute for the designs of trained professionals.

The pattern's sketches and colored patterns herein are free to use for everyone. Giving credit to the designer is not necessary but always be appreciated.

PREFACE

Purpose of this book is to share the Pattern Designs created by Mr. Muhammad Sheeraz S/o Abdul Hameed, who presented "Attainable Housing Standards" (The Affordable and Sustainable Housing).

This 1st Edition is comprising total 250 patterns 50 of them are the basic line works and remaining 200 are in different color shades using 50 line works. The color shades create different patterns. Some of them give illusion and some designs aim to transform the three-dimensional shape. These patterns can be used for; Architecture Massing, Architecture Surface Design (such as tile patterns, wall covering / wall cladding & etc.) Decoration, Textile Design, Art & Graphics, Branding, Logo Design, Packaging Design, Children Stuff *because patterns attract children* and many other purposes.

The more significant thing with these all patterns is that, all have been developed by initiating one line which followed by the concept of graph. The line of graph has been used by putting it with the different angles, with the combination of other lines, and in many patterns the graph line has been converted as arc line.

 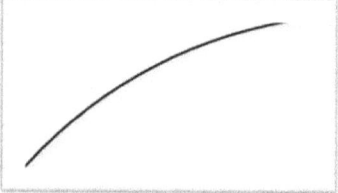

Pattern Designs

P-L-001

P-001-01

Pattern Designs by Sheeraz

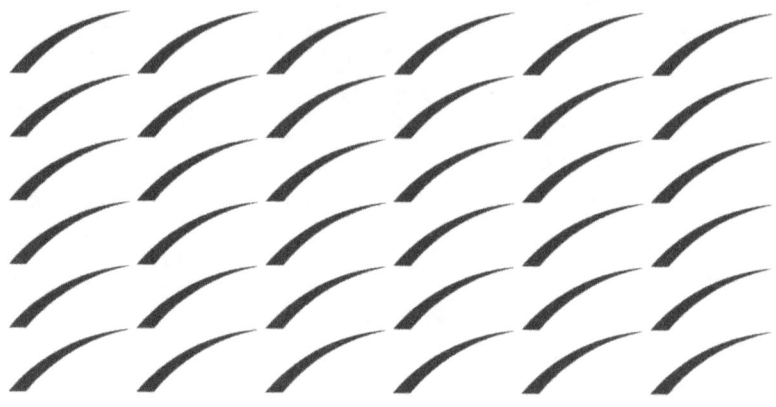

P-001-02

Pattern Designs by Sheeraz

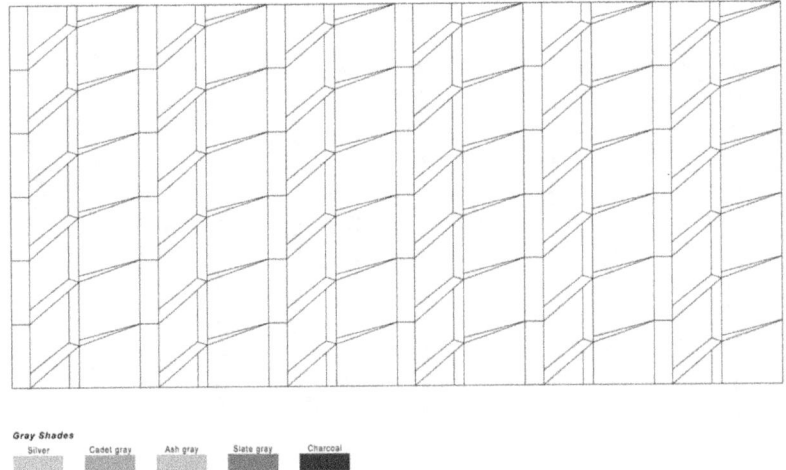

P-L-002

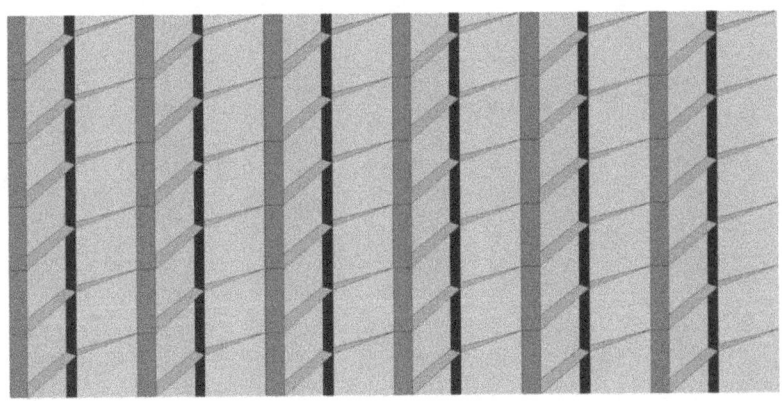

P-002-01

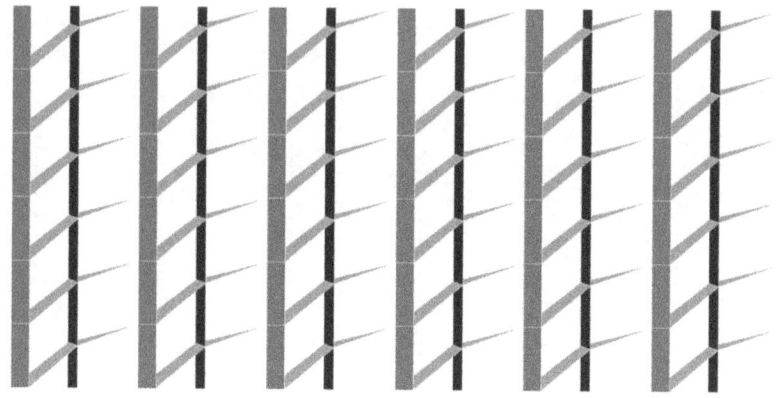

P-002-02

P-L-003

P-003-01

P-003-02

Pattern Designs by Sheeraz

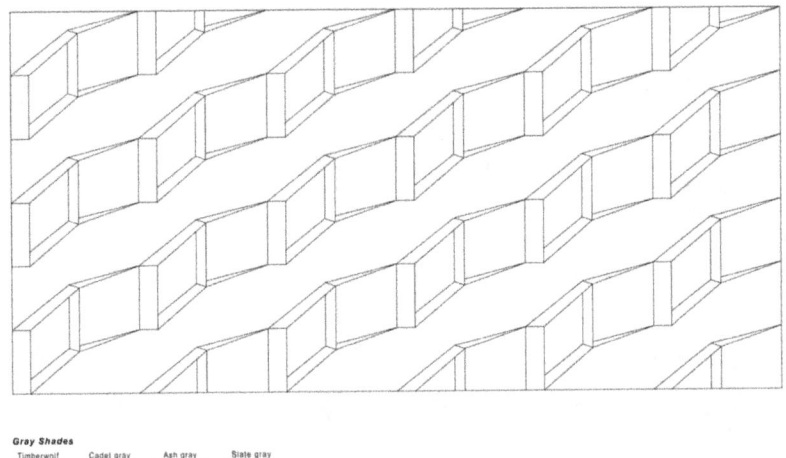

P-L-004

P-004-01

P-004-02

P-004-03

Pattern Designs by Sheeraz

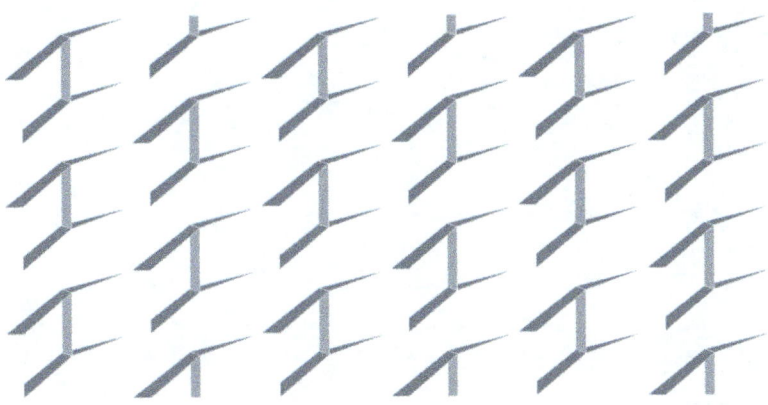

P-004-04

Pattern Designs by Sheeraz

P-L-005

P-005-01

P-005-02

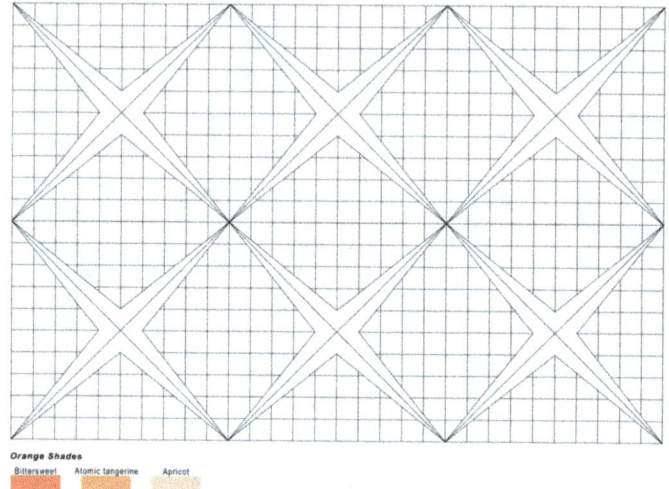

P-L-006

P-006-01

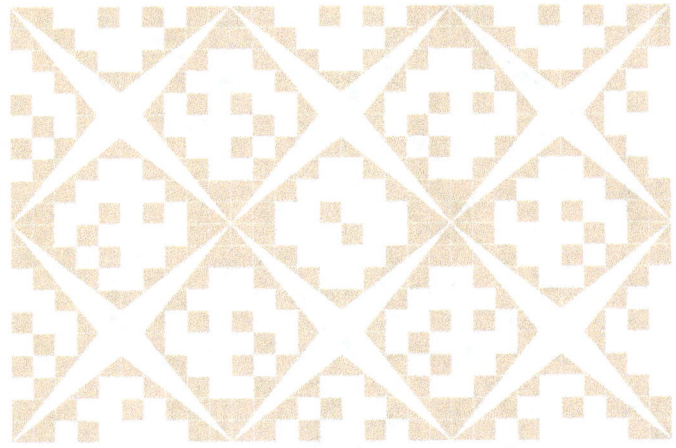

P-006-02

Pattern Designs by Sheeraz

P-L-007

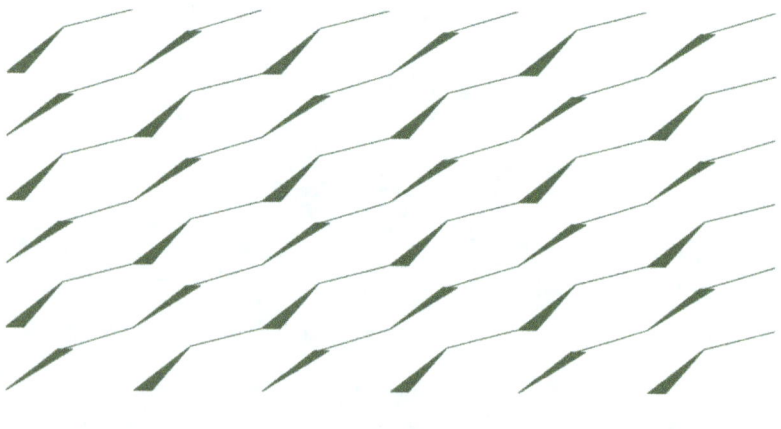

P-007-01

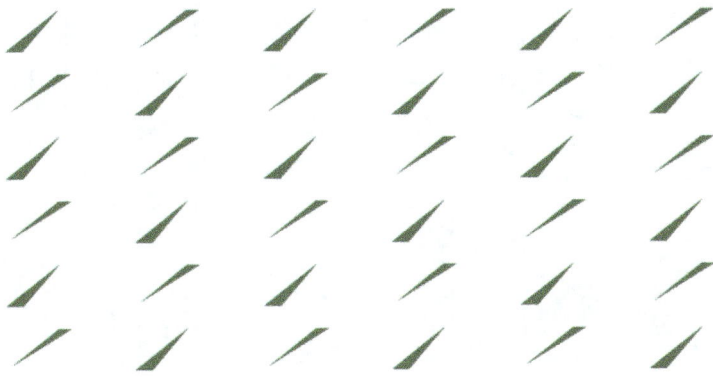

P-007-02

P-007-03

Pattern Designs by Sheeraz

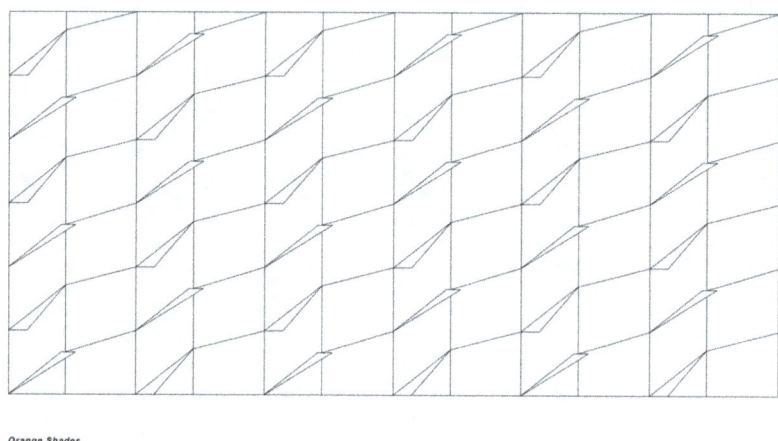

P-L-008

P-008-01

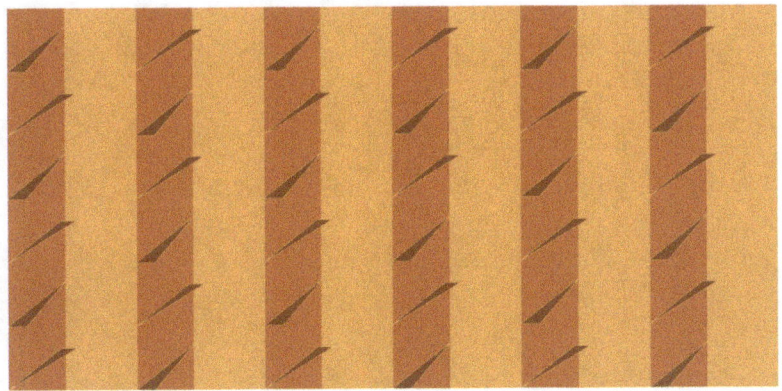

P-008-02

P-008-03

P-008-04

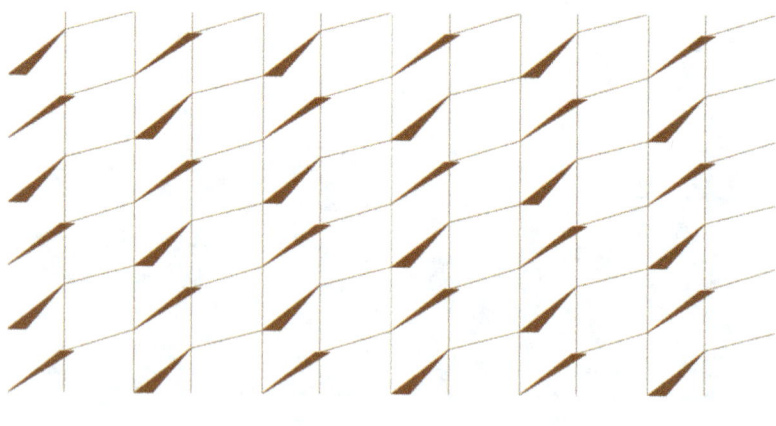

P-008-05

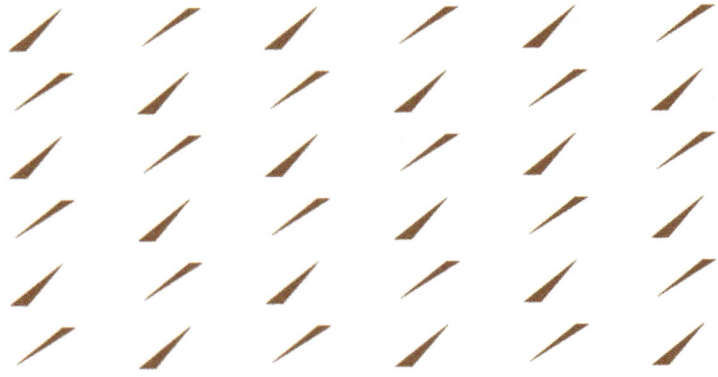

P-008-06

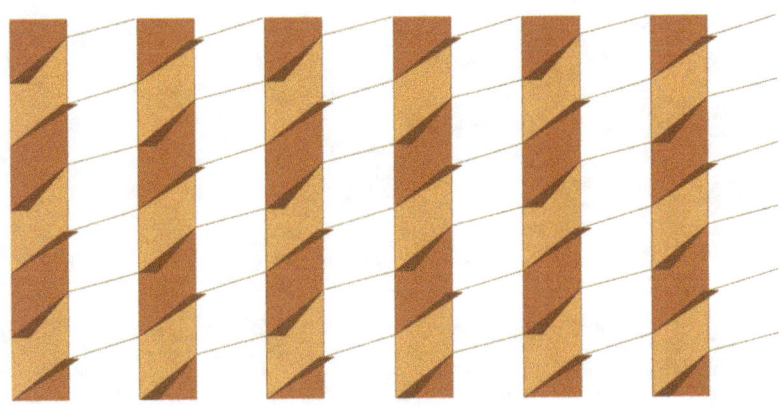

P-008-07

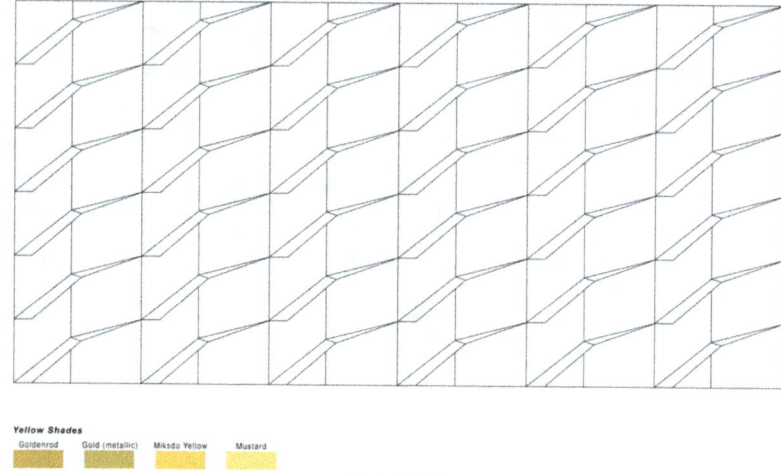

P-L-009

P-009-01

P-009-02

P-009-03

Pattern Designs by Sheeraz

P-009-04

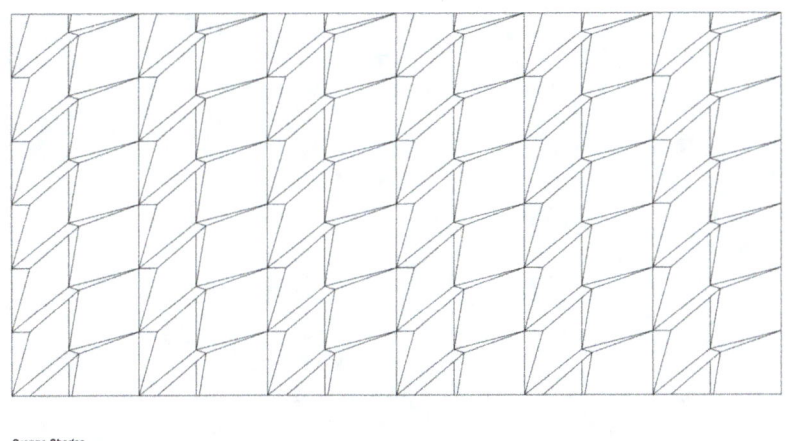

P-L-010

P-010-01

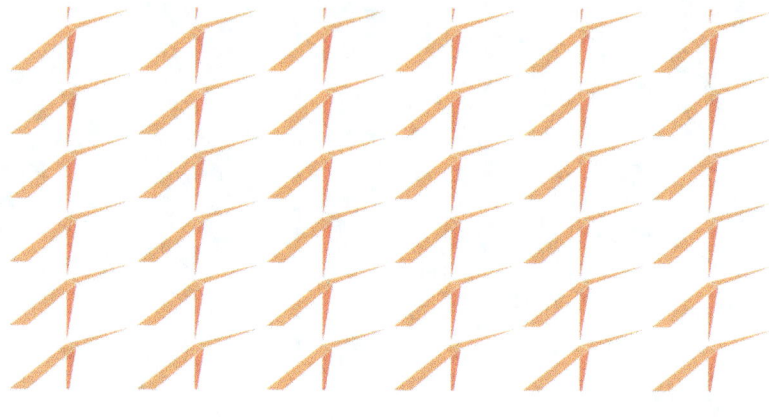

P-010-02

P-010-03

Pattern Designs by Sheeraz

P-010-04

Pattern Designs by Sheeraz

P-L-011

P-011-01

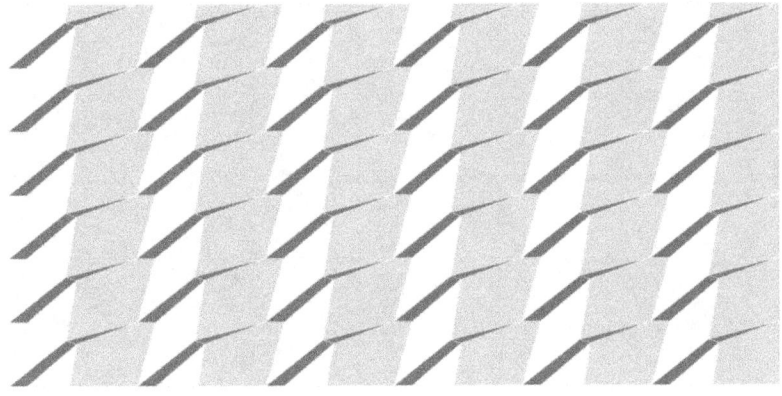

P-011-02

P-011-03

P-011-04

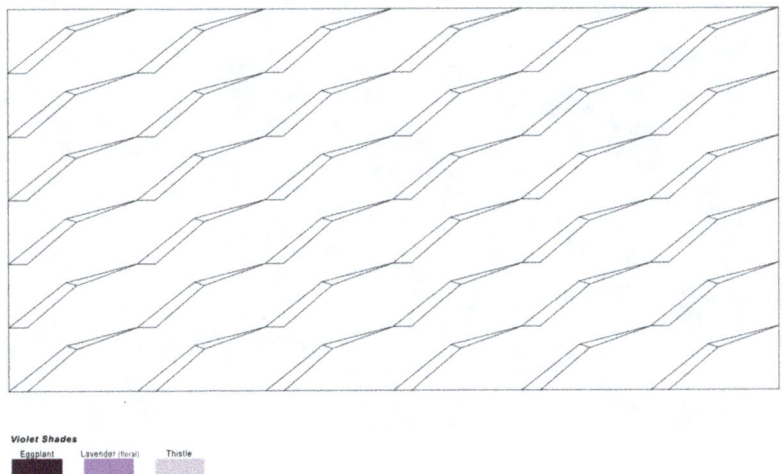

P-L-012

P-012-01

P-012-02

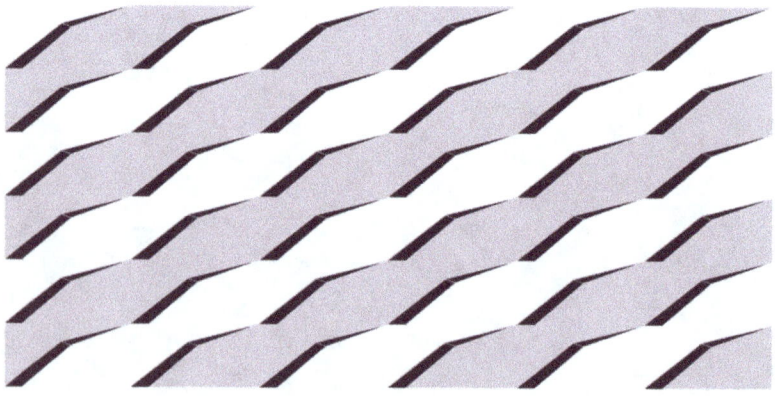

P-012-03

P-012-04

Pattern Designs by Sheeraz

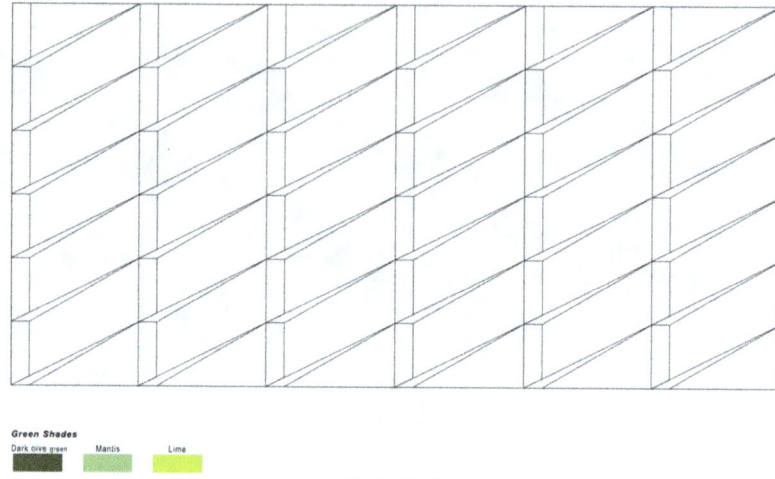

P-L-013

P-013-01

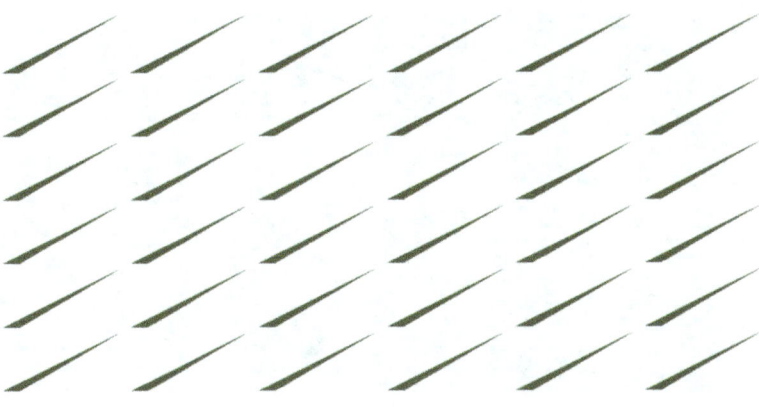

P-013-02

P-013-03

Pattern Designs by Sheeraz

P-013-04

P-013-05

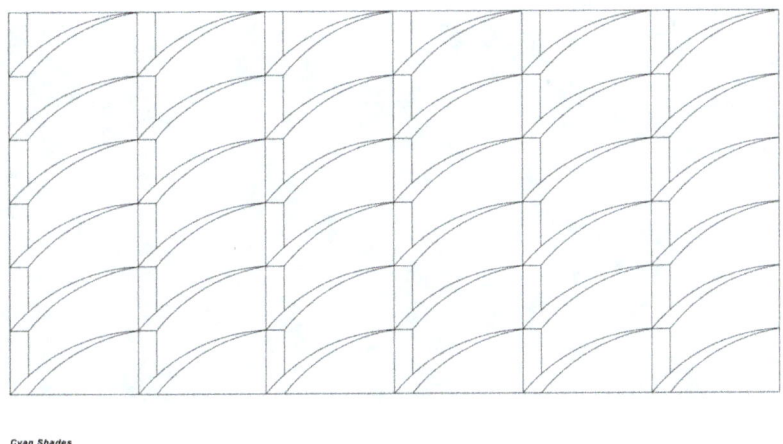

P-L-014

P-014-01

P-014-02

P-014-03

P-014-04

P-014-05

Pattern Designs by Sheeraz

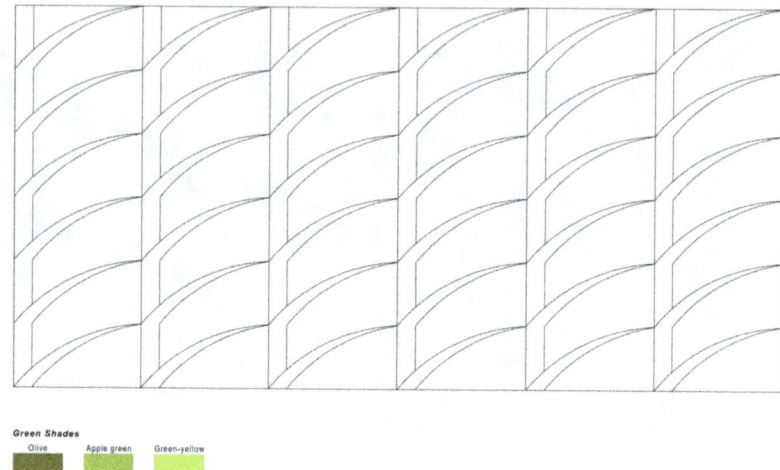

P-L-015

P-015-01

P-015-02

P-015-03

Pattern Designs by Sheeraz

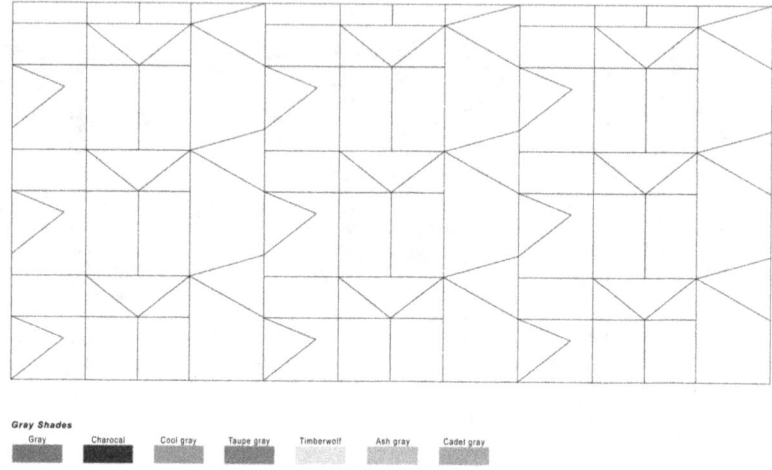

P-L-016

P-016-01

Pattern Designs by Sheeraz

P-016-02

P-016-03

P-016-04

P-016-05

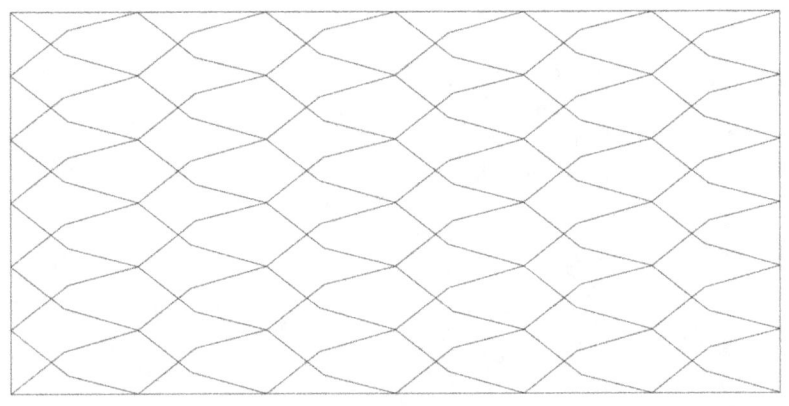

P-L-017

P-017-01

Pattern Designs by Sheeraz

P-017-02

P-017-03

P-017-04

P-017-05

Pattern Designs by Sheeraz

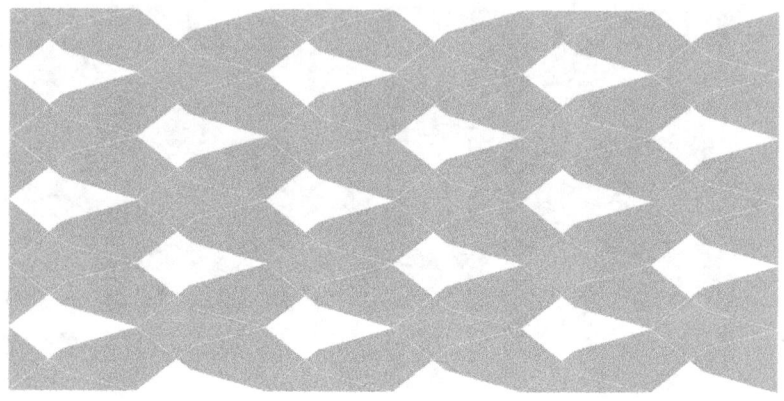

P-017-06

P-017-07

Pattern Designs by Sheeraz

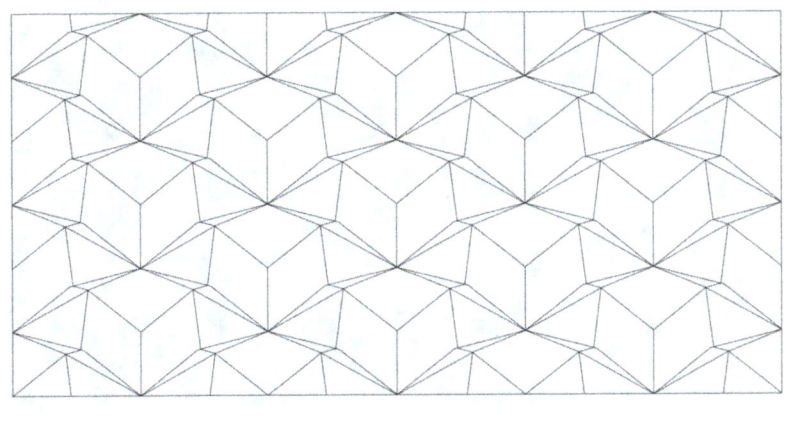

P-L-018

P-018-01

P-018-02

P-018-03

P-018-04

P-018-05

Pattern Designs by Sheeraz

P-018-06

P-018-07

P-018-08

P-018-09

P-018-10

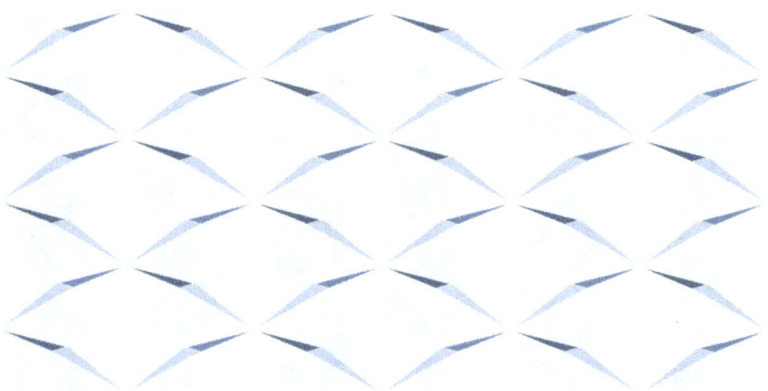

P-018-11

Pattern Designs by Sheeraz

P-L-019

P-019-01

Pattern Designs by Sheeraz

P-019-02

P-019-03

P-019-04

P-019-05

P-019-06

P-019-07

Pattern Designs by Sheeraz

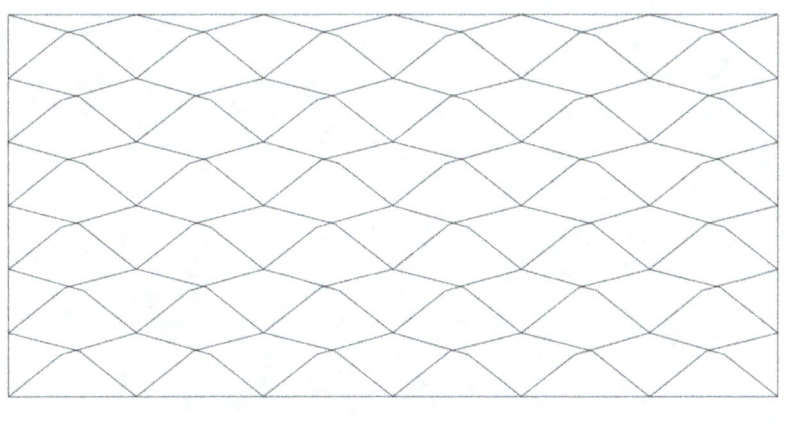

P-L-020

P-020-01

P-020-02

P-020-03

P-020-04

Pattern Designs by Sheeraz

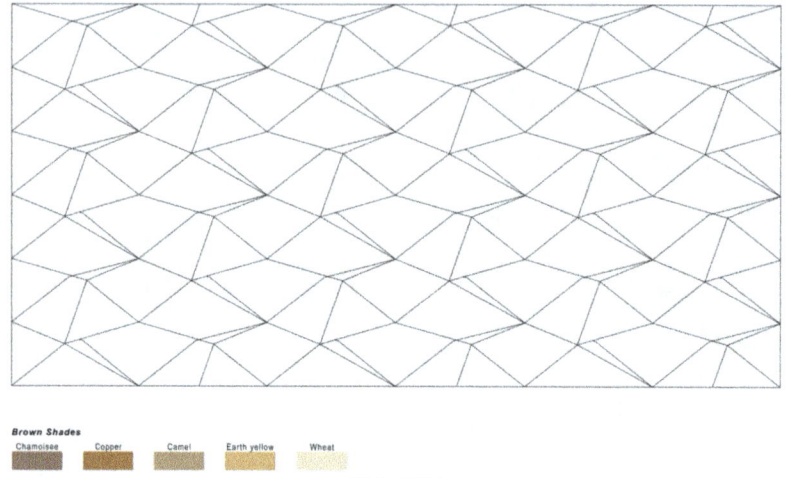

P-L-021

P-021-01

P-021-02

P-021-03

Pattern Designs by Sheeraz

P-021-04

P-021-05

P-021-06

P-021-07

Pattern Designs by Sheeraz

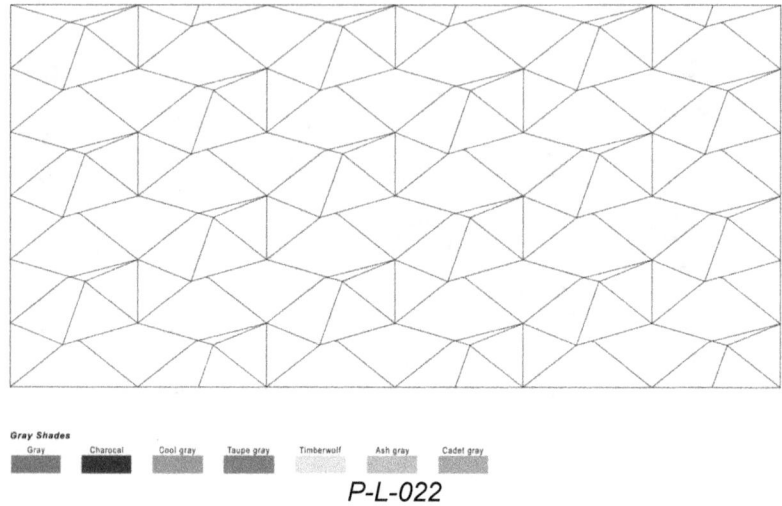

P-L-022

P-022-01

Pattern Designs by Sheeraz

P-022-02

P-022-03

P-022-04

P-022-05

Pattern Designs by Sheeraz

P-022-06

P-022-07

Pattern Designs by Sheeraz

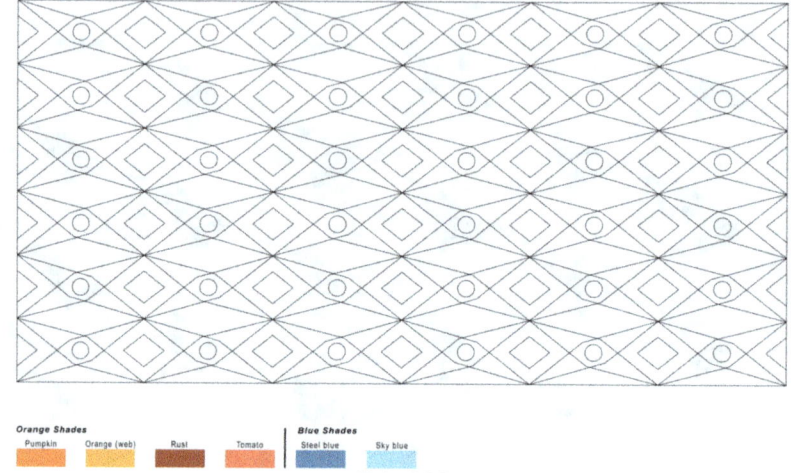

P-L-023

P-023-01

P-023-02

P-023-03

P-023-04

P-023-05

P-023-06

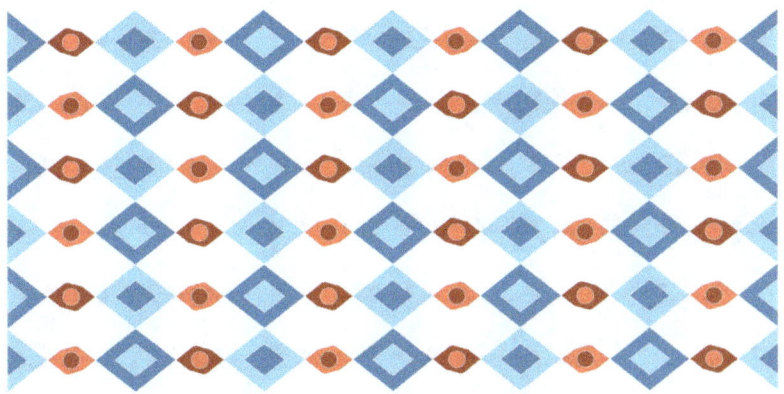

P-023-07

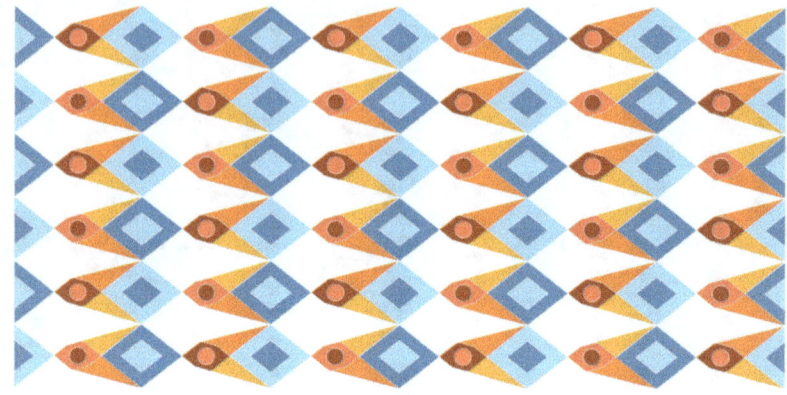

P-023-08

P-023-09

Pattern Designs by Sheeraz

P-023-10

Pattern Designs by Sheeraz

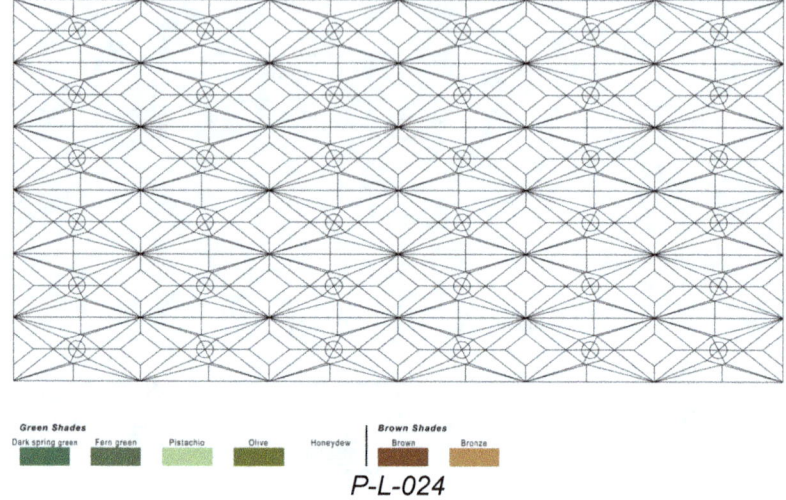

P-L-024

P-024-01

P-024-02

P-024-03

P-024-04

P-024-05

P-024-06

P-024-07

P-024-08

P-024-09

Pattern Designs by Sheeraz

P-024-10

P-L-025

P-025-01

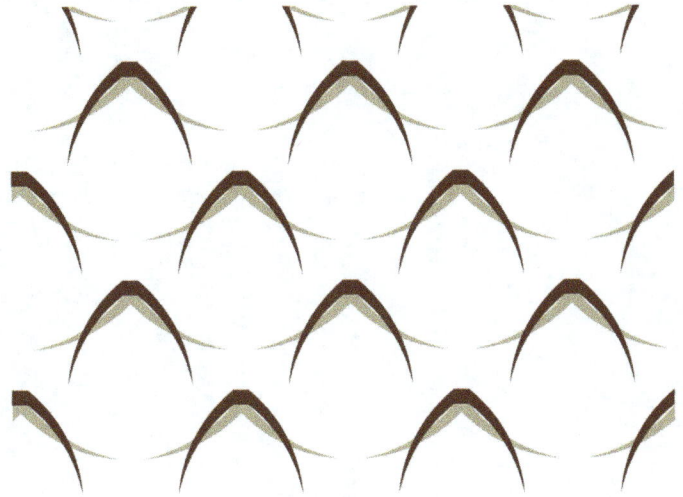

P-025-02

Pattern Designs by Sheeraz

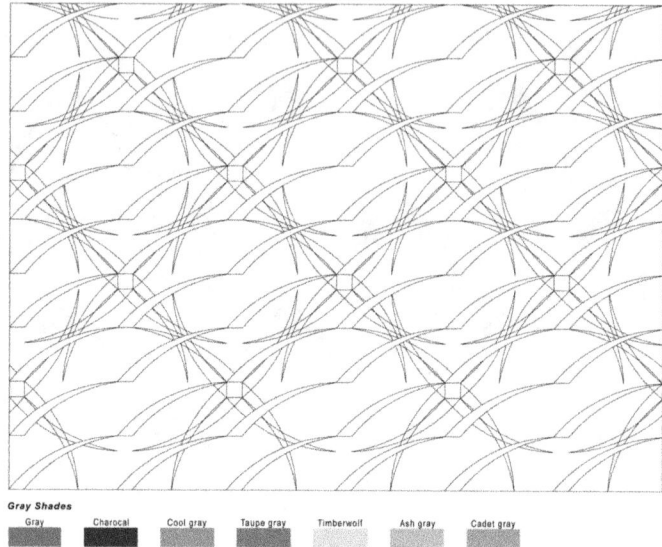

P-L-026

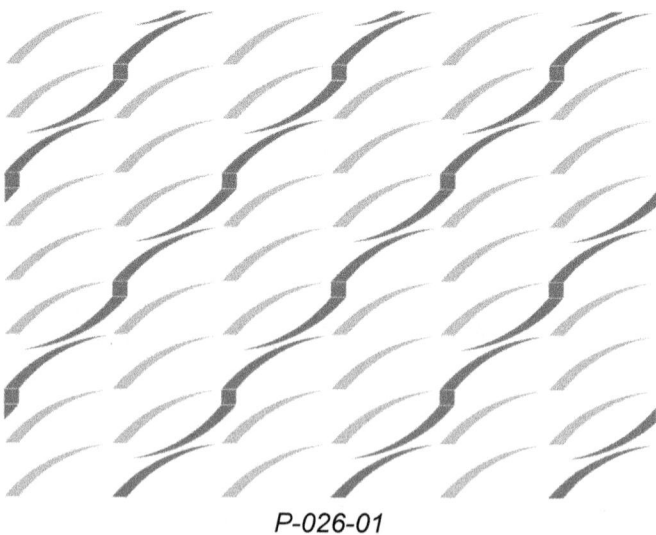

P-026-01

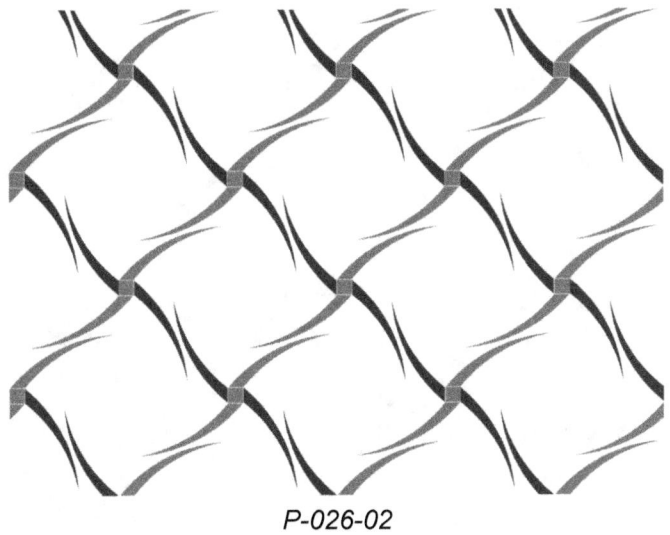
P-026-02

P-026-03

P-026-04

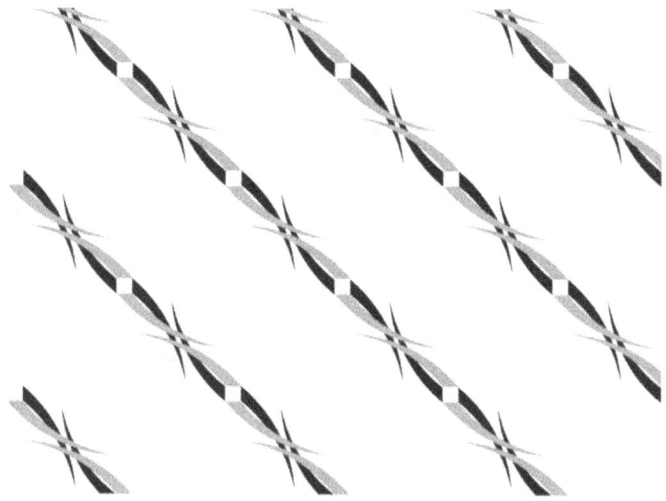

P-026-05

P-026-06

P-026-07

Pattern Designs by Sheeraz

P-L-027

P-027-01

P-027-02

P-027-03

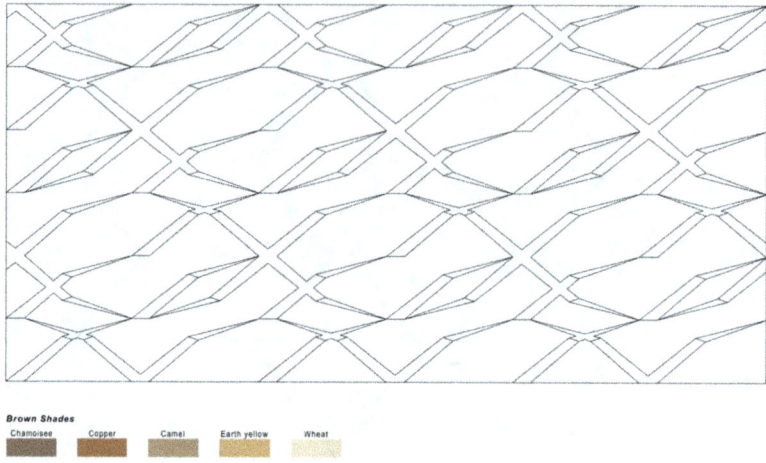

P-L-028

P-028-01

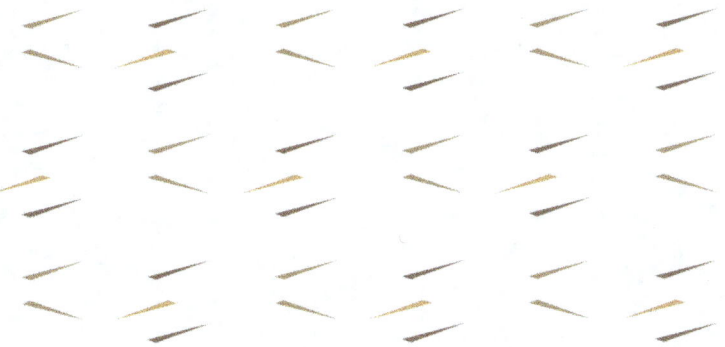

P-028-02

Pattern Designs by Sheeraz

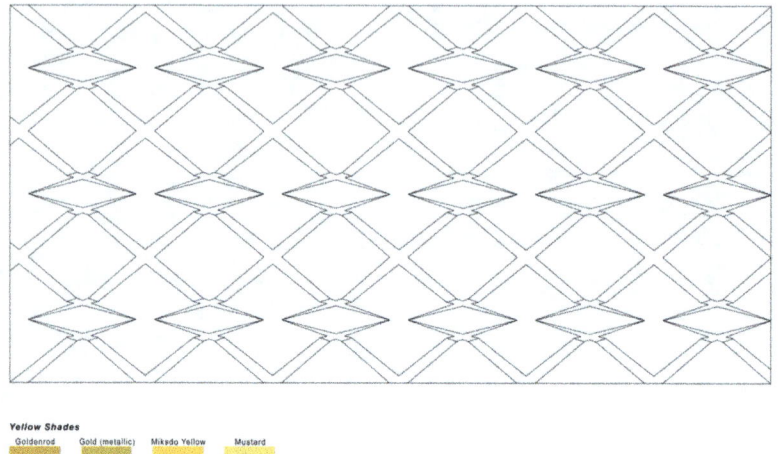

P-L-029

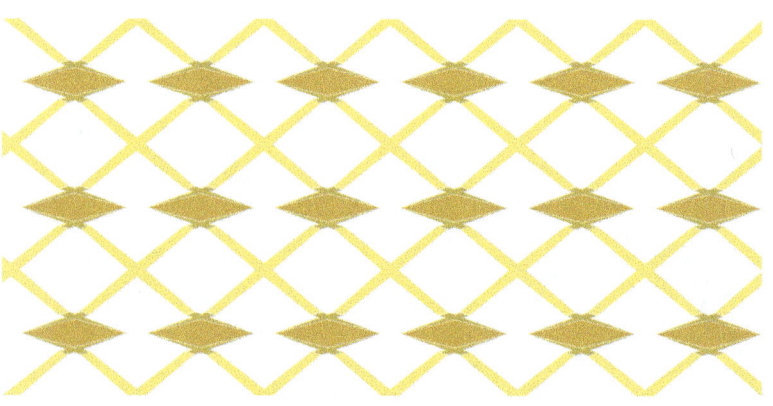

P-029-01

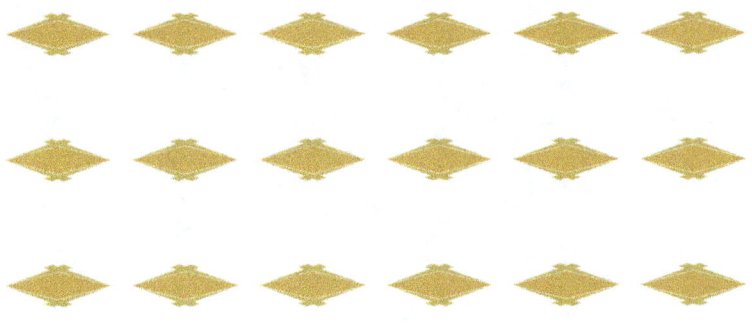

P-029-02

P-029-03

Pattern Designs by Sheeraz

P-L-030

P-030-01

P-030-02

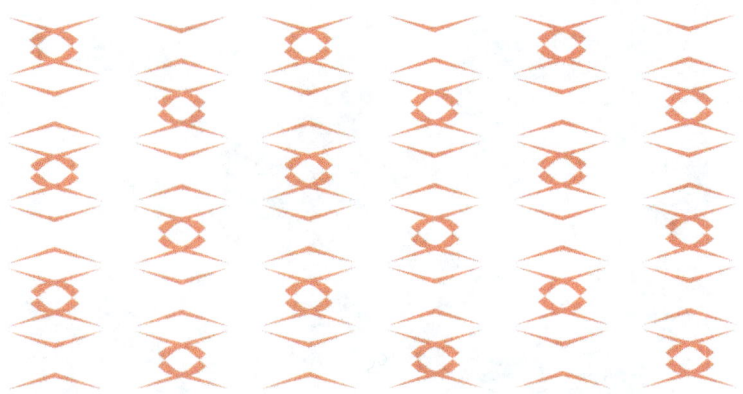

P-030-03

Pattern Designs by Sheeraz

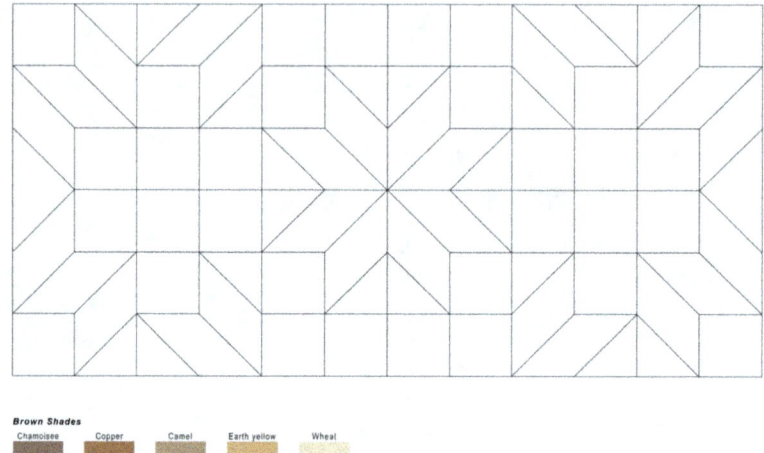

P-L-031

P-031-01

P-031-02

P-031-03

Pattern Designs by Sheeraz

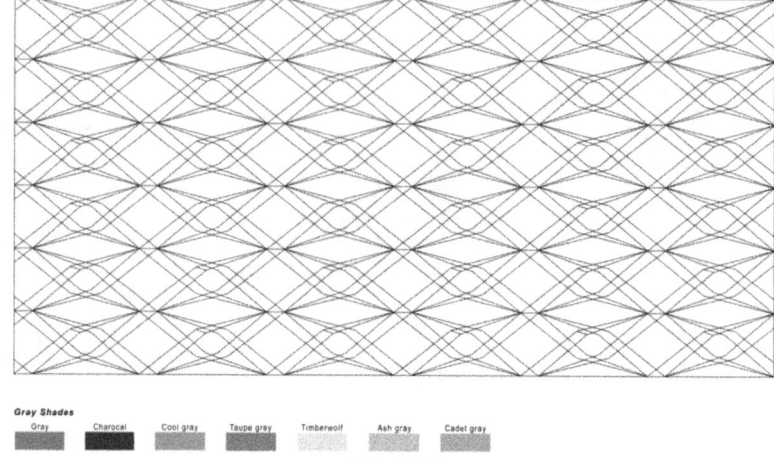

P-L-032

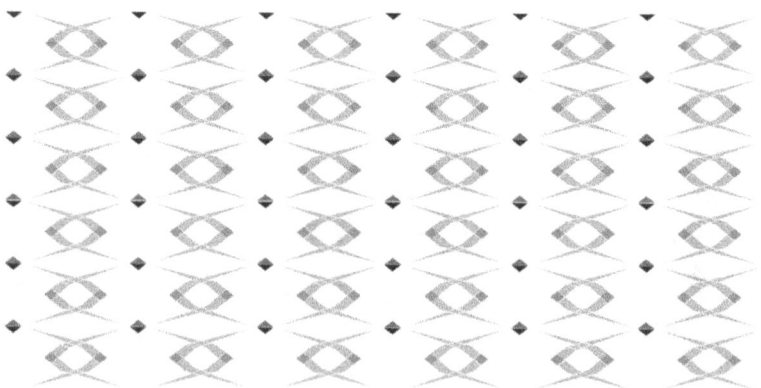

P-032-01

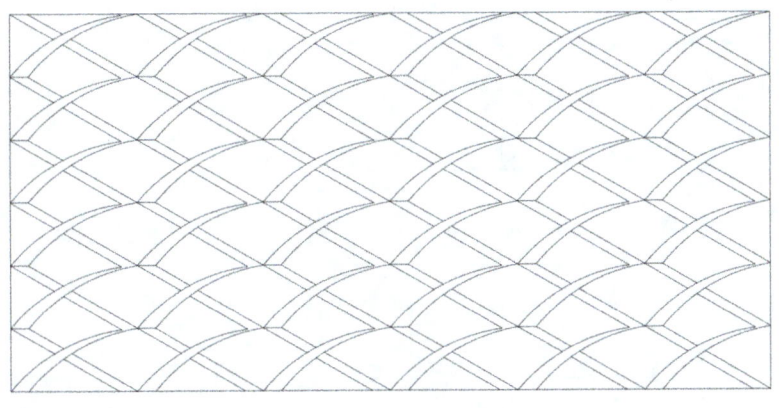

P-L-033

P-033-01

Pattern Designs by Sheeraz

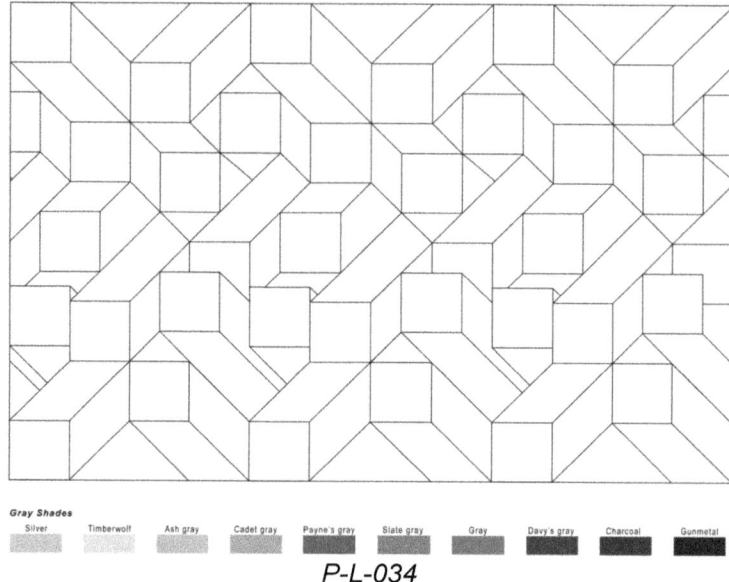

P-L-034

P-034-01

Pattern Designs by Sheeraz

P-034-02

P-034-03

P-034-04

P-034-05

P-034-06

Pattern Designs by Sheeraz

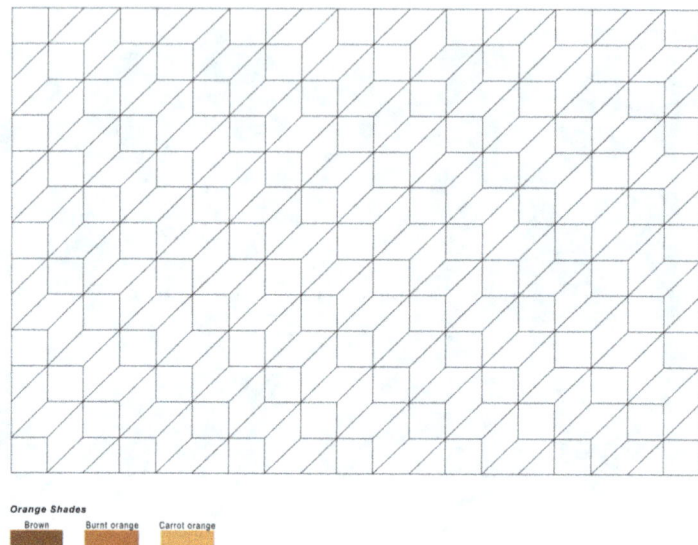

P-L-035

P-035-01

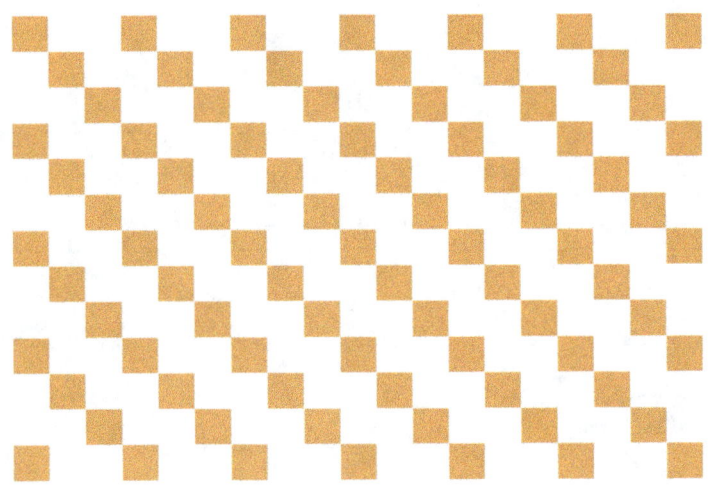

P-035-02

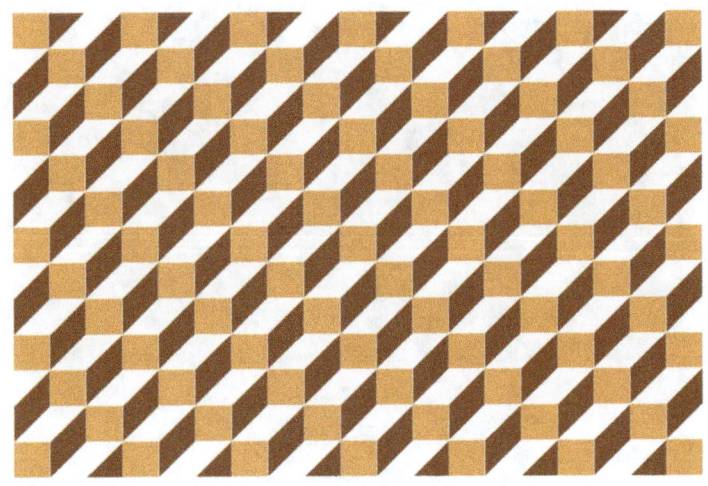

P-035-03

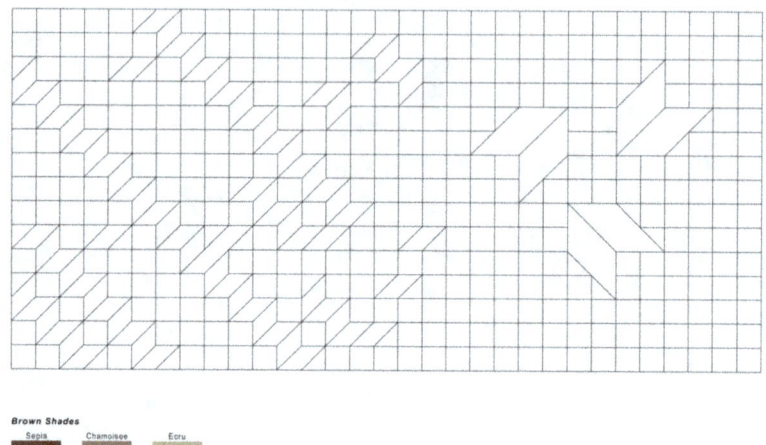

P-L-036

P-036-01

P-036-02

P-036-03

Pattern Designs by Sheeraz

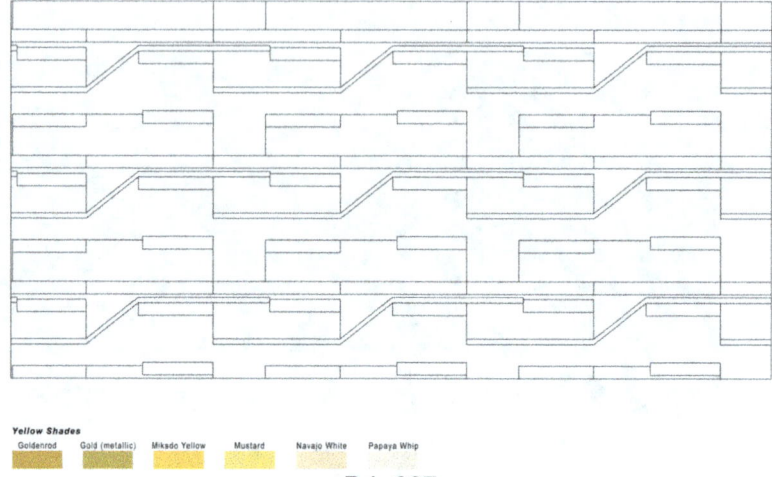

P-L-037

P-037-01

P-037-02

P-037-03

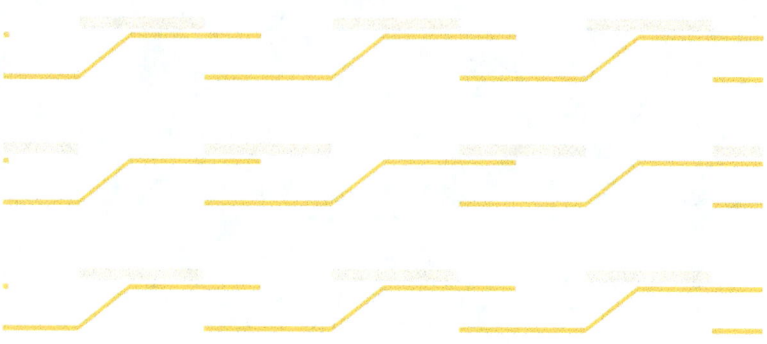

P-037-04

Pattern Designs by Sheeraz

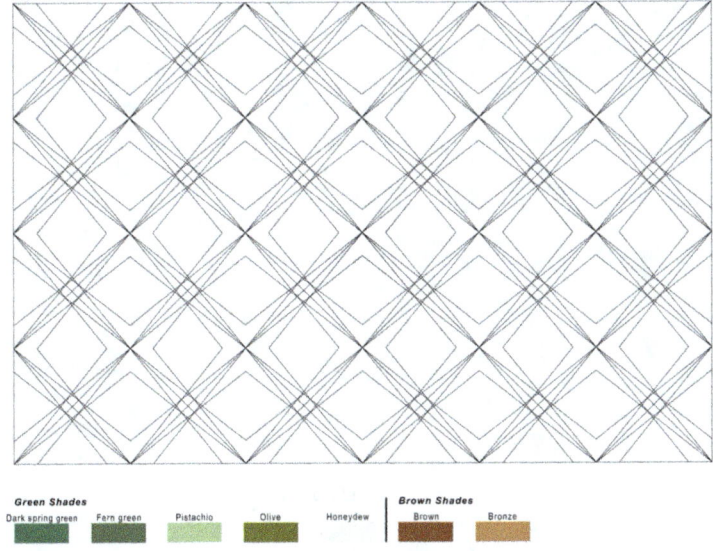

P-L-038

P-038-01

P-038-02

P-038-03

P-038-04

Pattern Designs by Sheeraz

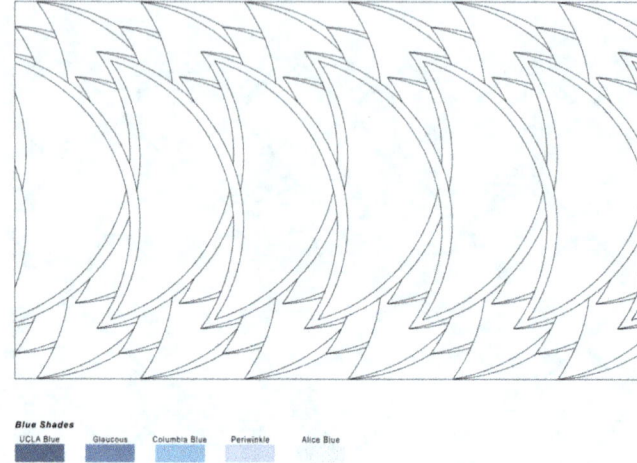

P-L-039

P-039-01

P-039-02

P-039-03

Pattern Designs by Sheeraz

P-L-040

P-040-01

P-040-02

P-040-03

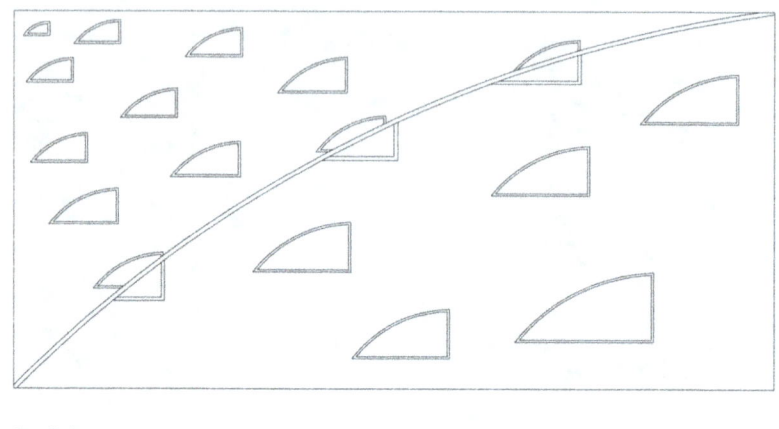

P-L-041

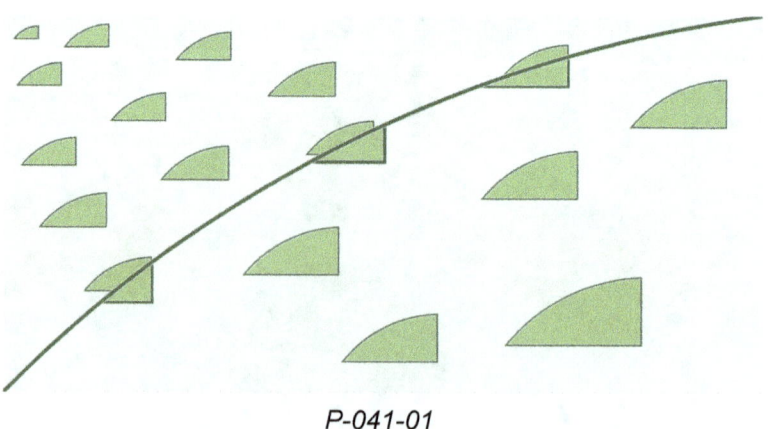

P-041-01

Pattern Designs by Sheeraz

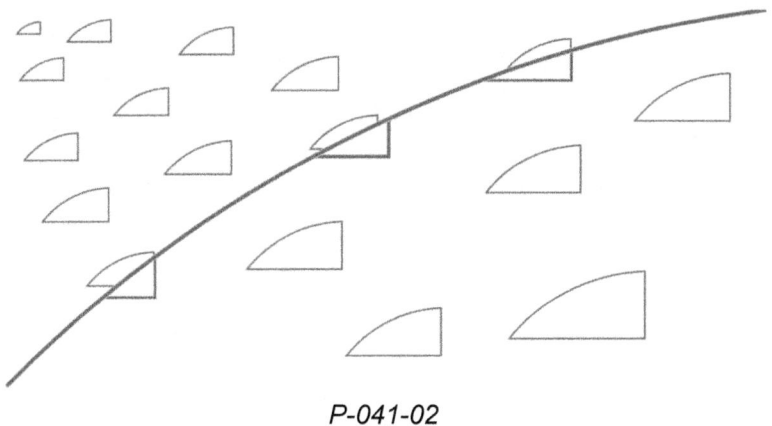

P-041-02

Pattern Designs by Sheeraz

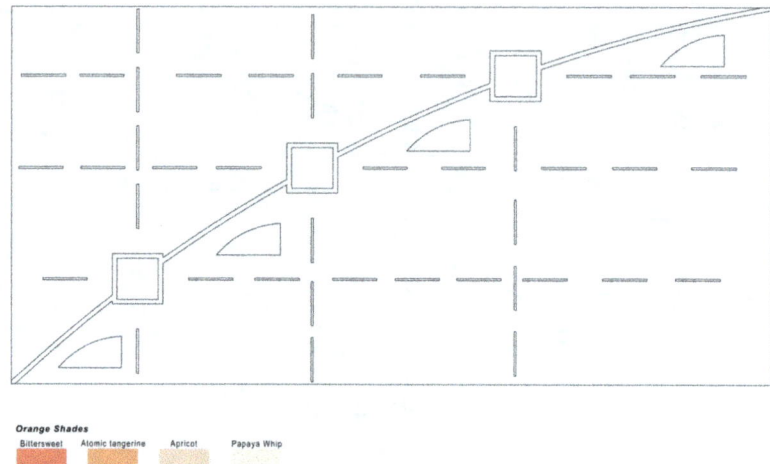

P-L-042

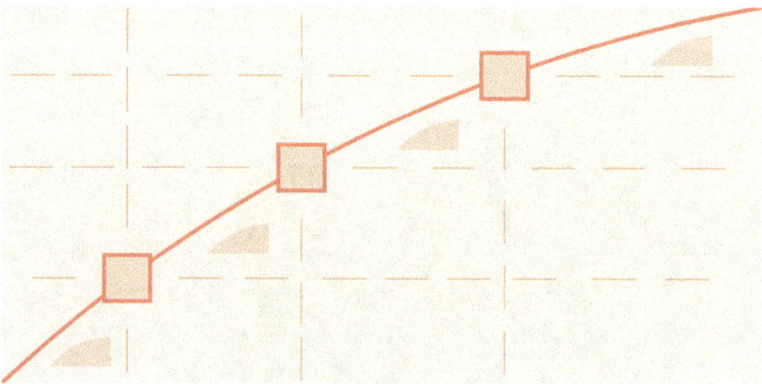

P-042-01

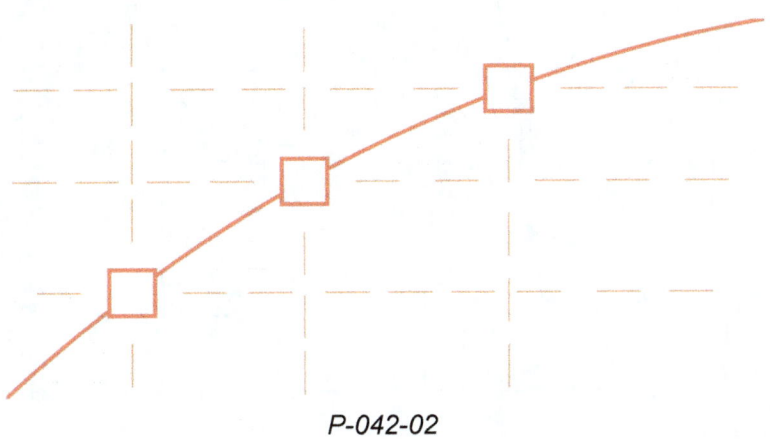
P-042-02

Pattern Designs by Sheeraz

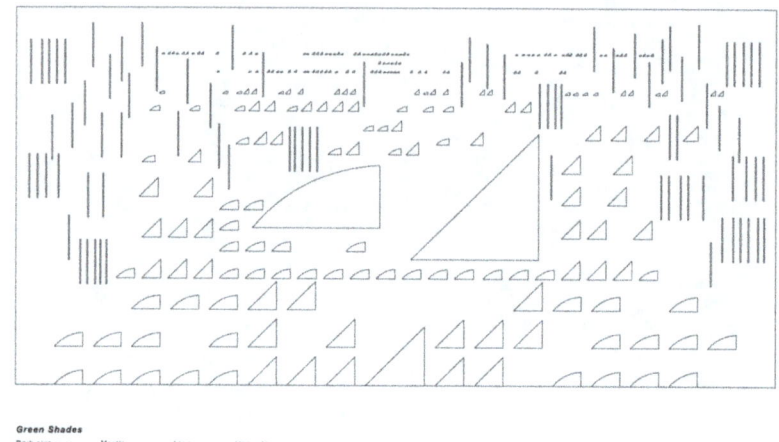

P-L-043

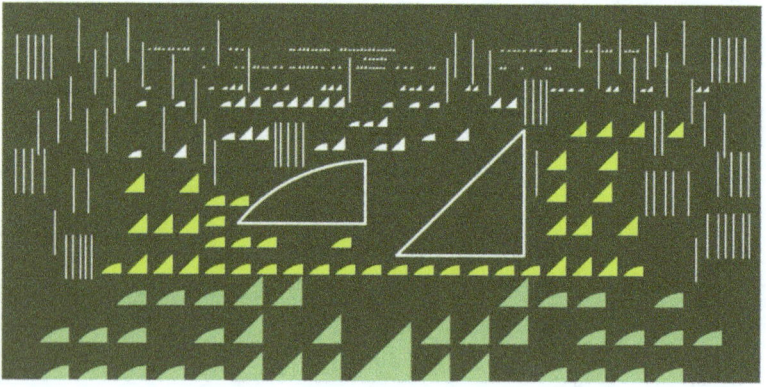

P-043-01

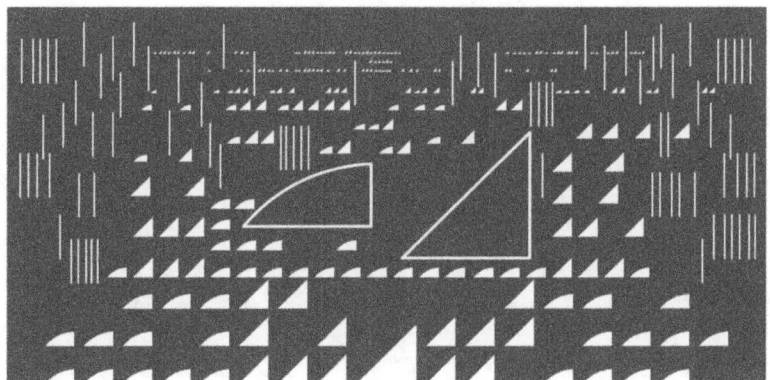

P-043-02

Pattern Designs by Sheeraz

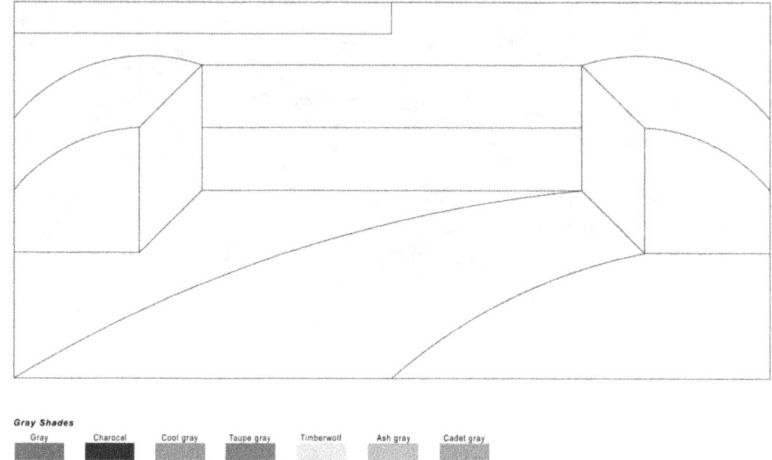

P-L-044

P-044-01

130

P-044-02

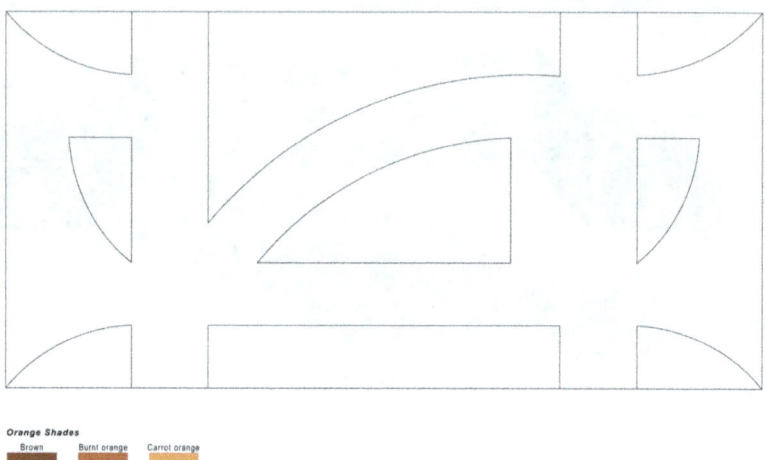

P-L-045

P-045-01

Pattern Designs by Sheeraz

P-045-02

P-045-03

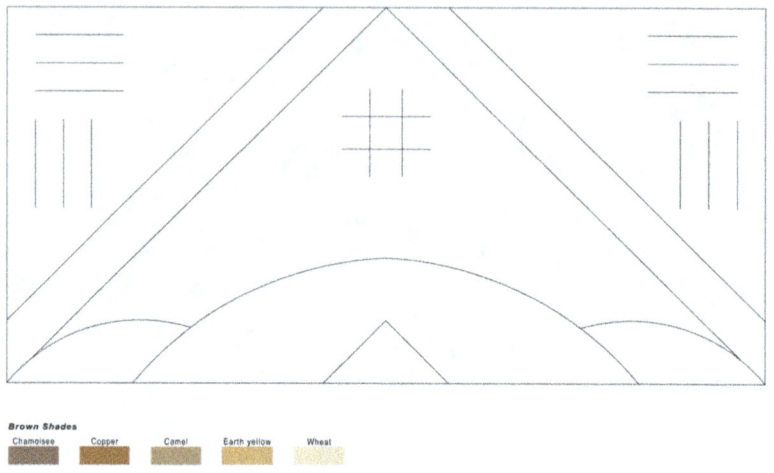

P-L-046

P-046-01

P-046-02

P-046-03

Pattern Designs by Sheeraz

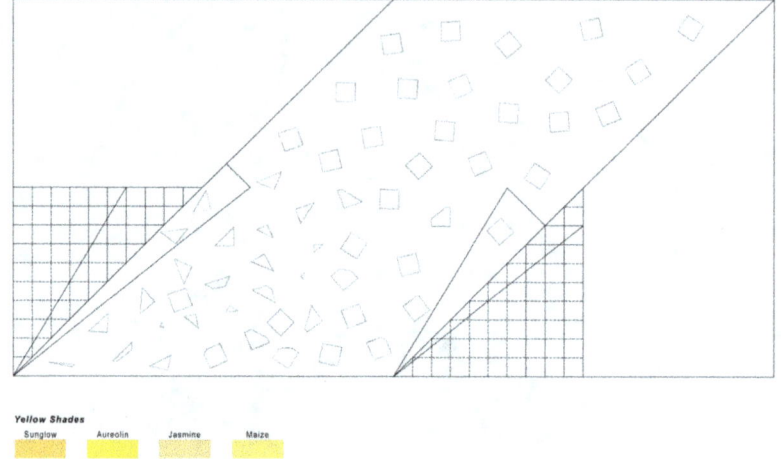

P-L-047

P-047-01

Pattern Designs by Sheeraz

P-047-02

P-047-03

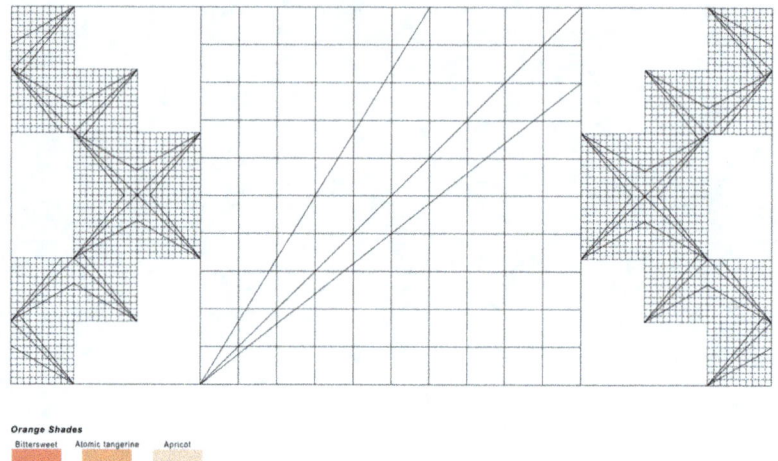

P-L-048

P-048-01

Pattern Designs by Sheeraz

P-048-02

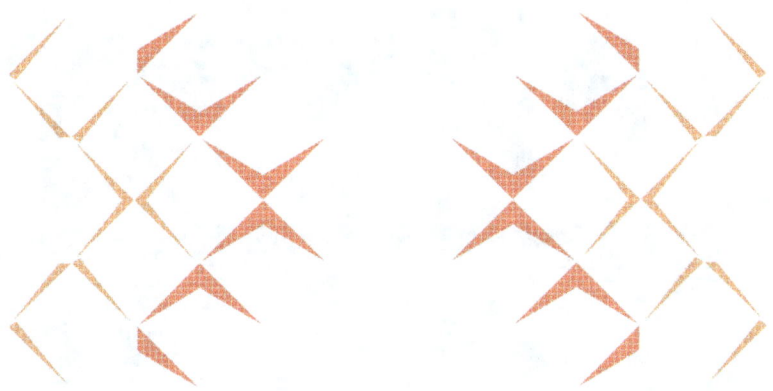

P-048-03

Pattern Designs by Sheeraz

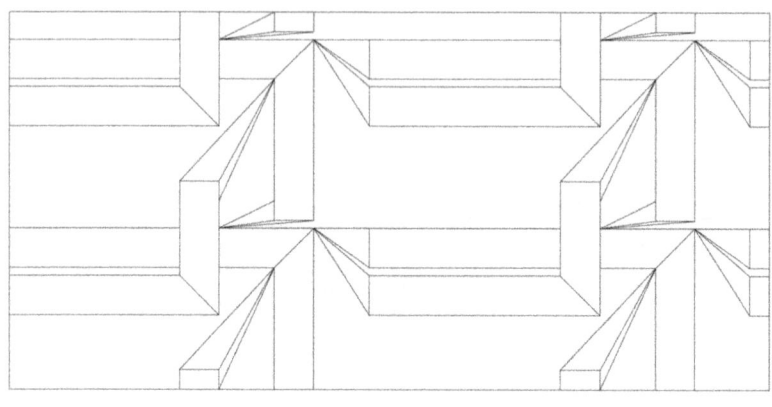

P-L-049

P-049-01

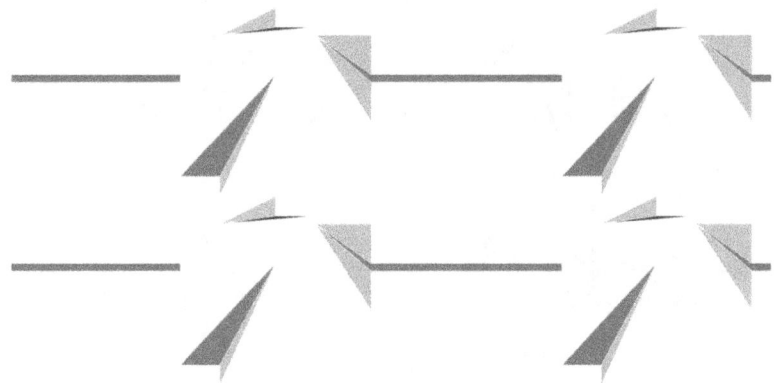

P-049-02

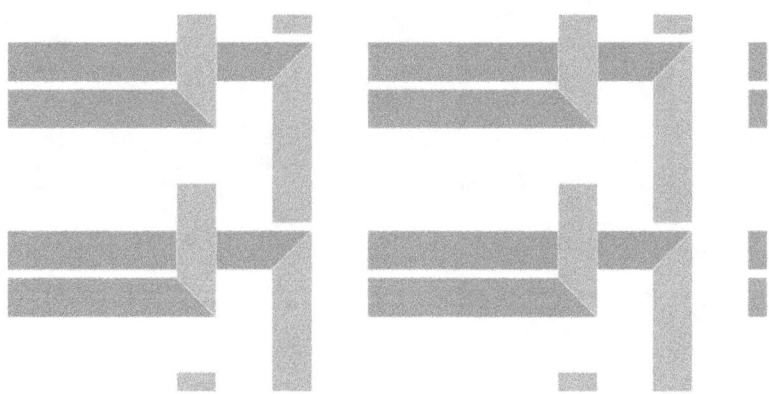

P-049-03

P-049-04

Pattern Designs by Sheeraz

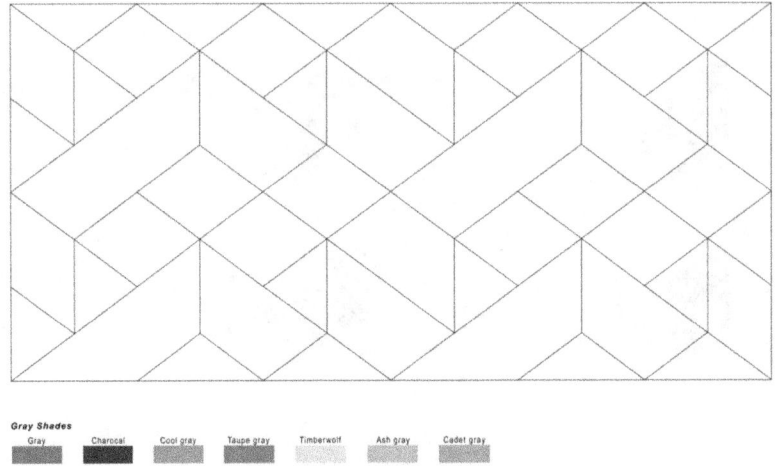

P-L-050

P-050-01

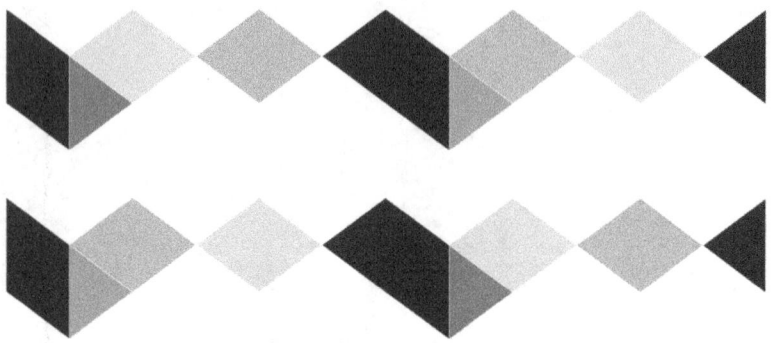

P-050-02

P-050-03

P-050-04

**Thanks for viewing!
Please provide your reviews
Your feedback is very important
Let me know what you thought!**

Reviews are extremely helpful, thank you for taking the time to support my work and me. Do not forget to share your reviews and encourage others to view:
"Pattern Designs by Sheeraz"

FOR DESIGNS,
NEW RELEASES
AND OTHER GREAT BOOKS
muhammadsheeraz26@gmail.com
Call & WhatsApp: +92 321 242 4862
www.ahspakistan.info

www.ingramcontent.com/pod-product-compliance
Lightning Source LLC
Chambersburg PA
CBHW060858170526
45158CB00001B/407